2,50

D1238326

# TROPICAL BLOSSOMS

## of the

# PACIFIC

Text and Color Photography

by

Dorothy and Bob Hargreaves

Copyright in Japan 1970/Dorothy and Bob Hargreaves
Printed in Japan

Published by
HARGREAVES COMPANY, INC.
Box 895 KAILUA, HAWAII, 96734 U.S.A.

## AFRICAN DAISY, Transvaal Daisy, Barberton Daisy

*Gerbera jamesonii*

A hairy perennial herb often grown for flower arranging. The flower stalks rise about a foot from the low growing leaves at their base supporting lovely 3″ to 4″ flowers. These large daisies are borne on a single stem. The colors range from pink to white to reddish-orange. There are single and double varieties. They are a native of Transvaal, Africa. Seen in Ceylon, New Caledonia, Hawaii and Philippines.

## ALLAMANDA, Large Yellow Bells, Golden Trumpet

- Hawaii: Lani-alii (Heavenly Chief) • Samoa: Pua-taunofo
- Thailand: Ban burie

*Allamanda cathartica*

These large, velvety, golden-yellow flowers grow on sprawling vines or shrubs. The tube spreads into five thick lobes with two or three buds opening at a time. The buds are pointed, brownish in color and look as if they had been varnished. The leaves are smooth, thick and pointed. The plant is of the Periwinkle Family from Brazil, and is often used as a cathartic —thus the botanical name.

## ALOCASIA, Elephant Ear

• Cook Islands, Rarotonga, Easter Isles: Kape • Fiji: Via, Via Mila, Via Sori • Guam: Piga • Hawaii, Tahiti, Samoa: Ape—pronounced "Ah-pay." • Malaysia: Keladi • New Caledonia: Wave, Pindu • Philippines: Biga (Tag.), Badiang • Ponape: Oht • Tonga: Ta'amu

*Alocasia macrorrhiza*

Very large, heart-shaped leaves 4′ to 5′ shelter these strange foot long flowers that have rather an unpleasant odor. They come from Asia and are all through Polynesia, Malaysia, New Hebrides, and the Philippines.

## AMARYLLIS, Barbados Lily

• Philippines: Orange Lirio • Thailand: Wan-see-tit

*Hippeastrum equestre*

These lovely large trumpet flowers can be seen throughout the Philippines, Fiji, S. E. Asia, and other S. Pacific Islands. Several flowers grow at the top of a 2′ stalk that rises from a brown-scaled bulb. There are many varieties with a number of spectacular color combinations.

## ANGEL'S TRUMPET TREE, Daturas, Angel's Tears

• Ceylon: Ratta-attana • Fiji: Davui, Boniwai • Hawaii: Nana Honua (gazing earthward) • Malaysia: Kechubong

*Datura candida*

The large swaying trumpets on this small tree or shrub look as if they had been placed there to decorate a Christmas Tree. They are around ten inches long and have five thin segments each coming to a twisted point. In the evening, they give off an exotic scent of musk. The leaves are large, greyish green, thick and velvety. The flowers and leaves are poisonous to eat, but the natives sometimes smoke them as an asthma remedy. The whole plant contains a strong narcotic. In Australia and New Zealand the plant can be cut to ground level and covered to protect it from frost.

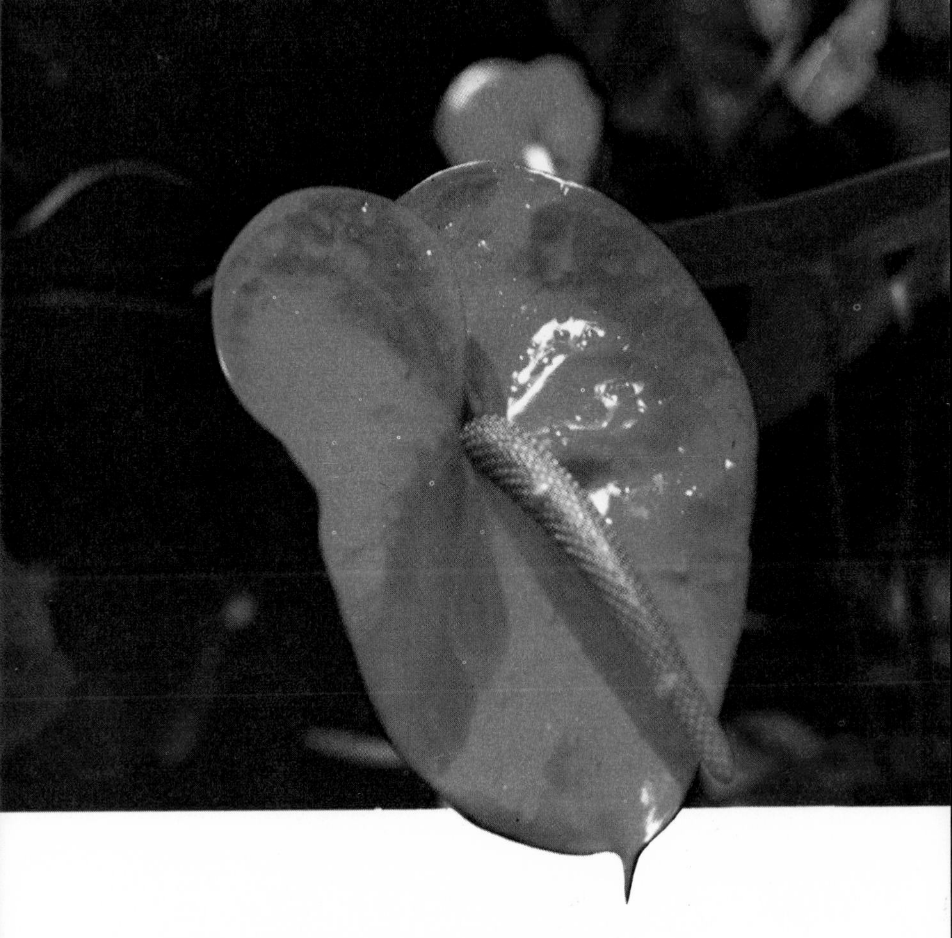

## ANTHURIUM, Flamingo Flower

• Thailand: Na-woa

*Anthurium andraeanum*

Like a mysterious unreal beauty, this waxen valentine truly typifies the tropical exotics. Ranging in color from pure white, pink, orange, deep red, green, they are very popular because of their long lasting quality, which is three weeks if cut in their prime.

The true flowers are on the center spadix and are hardly noticeable. It is the heart-shaped bract looking like oil cloth because of its thick and shiny appearance that is the eye catcher. Almost looking artificial in all its glorious beauty, the anthurium is indeed a favorite with all.

## PINK BIGNONIA, Mock Azalea, Desert Rose

• Thailand: Chuan Chom

*Adenium obesum*
syn. *A. coetaneum*

A smooth trunk and branched succulent shrub of 3′ to 6′ is from East Africa. The abundant milky sap is said to be poisonous. The pretty pink tubular flowers are about 2″ across. They make an interesting potted plant, as they are here around the Siam Intercontinental pool. Bangkok also has some of the plants in their medial strips. They thrive in dry locations. Some grow in Hawaii and other tropical areas.

## BIRD OF PARADISE

•Thailand: Prug-sa-sawan

*Strelitzia reginae*

The exotic coloring and looks of this flower is undoubtedly known to many. The tall stalks look like the neck of a bird topped by a lovely head with a long beak and crest. This beak is a pointed sheath, greyish-blue in color. The crest of the bird is made up of flowers lifting out of this sheath; usually about six to a sheath. One pushes out each day or so, thus the cluster becomes larger and more colorful as it becomes older. The flower has three pointed petals, brilliant orange with blue staminodiums shaped like arrowheads. The flower stalks grow slightly above the clump of stiff paddle-shaped leaves which are about three or four feet long with a reddish vein down the center. The plant is a relative of the banana and a native of South Africa. Plantings can be seen in the U.S. Cemetery in Manila, the Botanic Gardens in Sidney.

## OUGAINVILLEA, Paper Flower

• Bali: Kemong Katas (Paper flower). • Samoa: Fua pepa • Thailand: Fuang Fah • Tonga: Felila

*Bougainvillea glabra*

These paper-like flowers are long, colorful sprays of bright crimson bracts (the true flower is the tiny white flower in the center of the picture above), which are grown on a close cousin of the purple species (*B. spectabilis*) pictured right. This native of Brazil was named after a French navigator Louis A. de Bougainville. Many bright colored varieties may be seen all over the tropical world.

## BROMELIAD

*Bromelia humilis*

This plant has the ability to thrive on air as well as soil. The Bromeliads can often be seen perched on branches of trees and telephone wires. It is a relative of the pineapple as is the Spanish Moss one sees hanging from trees in the tropics. There are many of these epiphytes throughout the tropical world which are non-parasitic plants that grow on other plants, but get their nourishment from the air—somethimes called air plants.

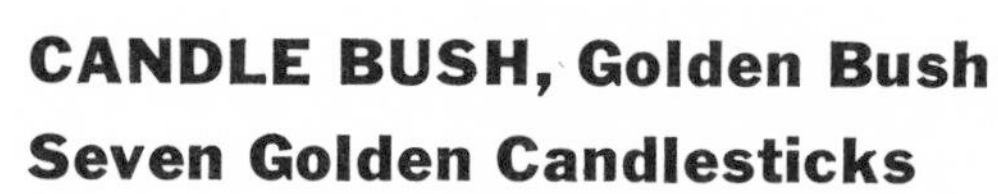

## CANDLE BUSH, Golden Bush
## Seven Golden Candlesticks

• Malaysia: Gelenggang • Philippines: Acapulco, Bikas (Tag.) • Tonga: Telango

*Cassia alata*

The closely packed spike heads of this bush look like yellow candles sticking up all over the plant. It is called God's Candle, or Roman Candle in Fiji. There are many blooming along the roads in Fiji, New Caledonia, and Ceylon. The bark is used for tanning, the leaves and seeds have medicinal value.

## CAPER BUSH

*Capparis zeylanica*

The Caper Bush is a spiny shrub with a 2″ orchid-like flower, white with a pink or yellow dab in it. It is used medicinally. The buds are sometimes pickled and eaten. From the flower buds of a species (*C. spinosa*) in the Mediterranean comes the "capers" of commerce. The long curved white stamens in the flower are sometimes referred to as the "Cat's Whiskers". Can be seen in the Medicinal Gardens of Royal Botanic Gardens in Ceylon.

There is a relative which grows on some S. Pacific islands called *C. sandwichiana*. In Hawaii this is called Pua-pilo, Maiapilo, which means "a pungent odor". This name comes from the bad smelling orange pulp of the 2″ green fruit. The flower itself is highly perfumed.

## CHENILLE PLANT, Monkey Tail, Pussy Tail

· Bali: Ikut Lutung · Samoa: Si'u-si'u Pusi · Thailand: Hang-Ga-Rok

*Acalypha hispida*

This strange looking tropical plant from the East Indies where it is used medicinally, has long velvety tails of dark red which resemble chenille. These tails are made up of staminate flowers with no petals. They have lovely dark green veined rather pointed leaves.

## CHEROKEE ROSE

• Hawai: Loke, Loke-hihi

*Rosa laevigata*

This single flowered fragrant rose of 2″ to 4″ in diameter grows wild all over the trees and bushes of some tropical islands. It is a persistent climber with smooth, shiny dark green leaves, and five white petals. Most roses are from the north temperate zone. This rose has been grown in North America for many years. The Chinese and the American Indians, from whom the name "Cherokee" comes, have similar legends explaining that roses have thorns to protect them. The rose is the National Emblem of England and the Flower of Maui, Hawaii. Can be seen blooming in Hong Kong, and Hawaii.

# CLERODENDRON, White Bleeding Heart, Glory Bower, Flower of Magic

• Ceylon: Ken-henda • Malaysia: Panggil Panggil, Bung Panngil, Sepanggil, Setawar • Philippines: Secreto de amor • Samoa: Afa, Mamalupe, Mamangi

*Clerodendron thomsonae*

Magical powers are attributed to this plant. In Malaysia they have been used medicinally for magic or curative powers—note names above indicating summoning spirits. The small white heart shaped clustered flowers have a bright red corolla with four tiny stamens that extend and seem to beckon, which according to Malaysians entice animals into a trap. This also gives the appearance of a bleeding heart. According to legend, a heart broken maiden, whose sweetheart had left her, wept tears and where they fell, a bleeding heart plant sprung up. The flower often thickens into an ornamental fruit that is a shiny red star. Some turn from green to purple to black.

## COCKSCOMB, Crested Cockscomb

• Hawaii: Lepe-a-moa

*Celosia argentea* var. *cristata*

This ornamental bright flower of the tropics is an annual herb with a peculiar shaped crested comb. The blossoms are of many colors—brilliant red, as in this picture, yellow, pink, orange and purple. They bloom on narrow spikes, plumed panickles, (*C. plumosa*), or comb-like crests. Can be see in the Singapore Botanic Gardens, Philippines, Hawaii, Hong Kong, Fiji and Ceylon.

## COFFEE, Arabian Coffee

- Ceylon: Kope, Kape, Caffeé, Kappé
- French: Caféier, Café
- Fiji: Kove
- Philippines: Kape
- Samoa: Kofe
- Tahiti: Taofe
- Thailand: Kafae

*Coffea arabica*

Coffee is one of the most important products exported from the tropics. It was first known in Abyssinia in the 15th century. Much coffee is grown in Java, Sumatra, Malaysia. Some is grown also in Ceylon, Thailand, Philippines, Fiji, Hawaii and other tropical islands. The coffee shrub is a small evergreen about 12 feet high with lovely shiny dark green leaves 3″ to 6″, and flowers that are very fragrant—similar to the Gardenia to which it is related.

## CONGEA, Shower of Orchids Vine.

*Congea tomentosa*

These mauve-pink velvety ramblers that spill over in a lovely starlike shower are really the bracts practically hiding the tiny white flowers in the center. These bracts, covered by woolly hairs underneath, are long lasting and make graceful flower arrangements. The vine is a native of Malaysia and Burma related to the Petrea which it resembles (see page 53). Grows in Fiji, Philippines, Hawaii. Can be seen in the Royal Botanic Gardens, Ceylon, and Waterfall Gardens, Penang, Malaysia.

## COTTON PLANT, Sea-island Cotton

· Ceylon: Pulun · Hawaii: Pulupulu-Haole

*Gossypium barbadense*

The seeds of this small shrub are formed from the yellowish flowers seen in the picture. These seed cases or bolls contain seeds wrapped in a mass of white cotton fibers. They can be used for oil, fertilizer, stock feed, soap, oilcloth, putty and nitroglycerine. The fibers, of course, produce the "Cotton" of commercial fame.

In India when the first cotton boll bursts open, it is saluted as "Mother Cotton" and worshiped in the hope that the crop will be profuse. There is a Philippine legend about a mother taking one of her twin boys to the cotton fields and laying him on some cotton fiber, but the wind carried the baby and cotton far away. He grew up to be a great warrior. The other twin also became a warrior. Not knowing that they were brothers, they met and fell in love with the same woman, and the rivalry was settled only when they discovered that they were twins. Grown in Ceylon, Asia, Cambodia, New Zealand, Thailand, S. Pacific.

## CROWN FLOWER, Indian Milkweed, White Ivory Plant, Asclepiad Tree

- Hawaii: Pua Kalaunu • Malaysia: Remigu, Remiga, Merigu • Philippines: Kapal-Kapal (Tag.) • Samoa: Pepe • Thailand: Dok-ruck

*Calotropis gigantea*

This Indian milkweed grows wild in many areas of the Pacific and S. E. Asia. It is used in Thailand combined with rose petals and jasmine as a garland when one leaves or arrives (the Hawaiians use it for a lei—see right.) Also in Thailand most homes have a Spirit House where Chao Ti lives. Offerings often include these flowers. There is also a purple variety (see right).

# FLY CATCHING PLANT, Pitcher plant

• Ceylon: Bandura-wel • Malaysia: Periok Kera • Philippines: Opli, Aran-kalau, Batidor, Gorgorita, Inomang kalau.

*Nepenthes gobmouche*

This curious climbing plant from New Caledonia is one of the many Pitcher Plants that grow from Malaysia to Ceylon.

Insects are attracted by the honey glands within these receptacles. They venture inside and slip down into water at the bottom, where they soon drown, and the plant then derives food from their decayed bodies through a digestive fluid secreted by the "pitchers" (similar to the digestive pepsin in the stomach). The flowers of this unusual plant are inconspicuous, but the pitcher is formed from a prolongation of the midrib which has a lid at the top that closes when several insects have been caught and the amazing digestive process begins. Grows also in Borneo, the Philippines, Singapore, Hong Kong.

# FRANGIPANI, Pagoda Tree, Temple Flower, Plumeria

• Bali: Kambodia, Djepun • Ceylon: Aralyna Pansal Mal • Fiji: Bua ni Vavalagi • Hawaii: Pua Melia • Philippines: Kalachuchi • Samoa: Pua Fiti • Thailand: Lan Tome • Tonga: Kalosepani

*Plumeria obtusa*

This lovely big white blossom is called the Singapore Plumeria. It usually keeps its large dark green Rhododendron-like leaves the year around, while the others shed theirs. There are many colors and varieties of these fragrant flowers. Because of their wonderful sweetness of scent, they are often planted near temples and burying grounds in Ceylon, S. E. Asia, Hawaii and other Pacific tropics. Thus the name Temple Flower.

## GARDENIA

- Hawaii: Keile
- Philippines: Rosal
- Samoa: Pua fua tausaga

*Gardenia jasminoides*

This handsome highly fragrant flower from China is a favorite the world over. Double ones look like a large, white rose. The blossoms are used for leis, corsages and flower arrangements. The shrub also has shiny dark attractive evergreen leaves. Even the fruits are attractive, changing from green to orange or red. A dye can be made from them.

## TIARE

*Gardenia taitensis*

This is the flower of Tahiti. They continually blossom on this attractive shrub from the Society Islands, and are also very fragrant. They are used for leis and head garlands in Tahiti and other Islands. Their essence is combined in coconut oil for the skin and hair.

## GINGER

### Red Ginger

- Samoa: ‘Avapuhi-fua-mumu
- Thailand: Khing-Dang

*Alpinia purpurata*

The long pretty waxen red bracts of these showy flowers look like the bloom, but the true flower is the small white blossom in the picture above. The plant is a native of Malaysia.

### Torch Ginger

- Samoa: ‘Ava Puhi, ‘Ava Puho

*Phaeomeria magnifica*

These are one of the showy heads of the world! The 15′ bamboo-like stalks and large green leaves are not unlike a small forest. They practically conceal the spectacular torch bloom that springs up independently 6’ tall.

### Shell Ginger

- Bali: Djahé • Fiji: Locoloco • Ceylon: Rankiriya • Philippines: Lankausas Napula

*Alpinia nutans*

The shell-like flowers which spill out of the top of this 12′ stalk have a porcelain look. Native of E. Asia. Prolific bloomers.

# GOLDEN DEW DROP, Duranta, Pidgeonberry

*Duranta repens*

A 6′ to 18′ evergreen shrub which flowers and fruits throughout the year. The flowers are tubular, about $\frac{1}{2}''$ wide, lilac-blue with violet stripes (also a white variety). The golden berries hang on the tree a long time in graceful trusses, and give the tree its common name. It has been popular in the Philippines since 1880. Many in Thailand, Bali, Australia, Hawaii. Can be seen at the Bogor Gardens in Indonesia.

# HELICONIA, Lobster Claw, Crab Claw, False Bird of Paradise

• Thailand: Kampoo • Tonga: Kavapui

*Heliconia* sp. 1—formerly *Heliconia humilis*

Another member of the Banana Family, is this bright red bract that suggest the red of boiled lobster claws. They grow in a clump of tall paddle shaped leaves. (The Hanging Heliconia on the cover is *H. rostrata*)

## HIBISCUS

• Bali: Kembang Sepatu • Fiji: Kauti, Lolonu, Senicikobia, Senitoa • Malaysia: Bunga raya • Tahiti: 'Aute • Philippines: Gumaméla • Thailand: Chabá • Samoa: 'Aute, Fautu • Tonga: Kaute

The Hibiscus is the outstanding flower of the tropics. As crossing is easy, there are many species of Hybrid Hibiscus. Many hybrids have been sent from Hawaii where it is the 50th State Flower. It is also the National Flower of Fiji and Malaysia. Women wear them in their hair. Men wear them behind their ears. They are unique as they do not wilt for a day out of water, but at the close of the day, they fade away after giving the world a beautiful flower for the brief span of a day. Throughout the world they furnish beauty, food (Okra), medicine, perfume and dye.

## HIBISCUS (continued)

### Tree Hibiscus

• Fiji: Vau • French: Bourao • Guam: Pago • Hawaii: Hau • Hong Kong: Sha Tin • Malaysia: Baru Baru • Philippines: Balibago, Malabago (Tag.) • Samoa, Tahiti, Tonga: Fau

*Hibiscus tiliaceus*

This is a true hibiscus. Often used for arbors. The bark is used for caulking boats and cordage. The Tahitians put flowers on it for head garlands.

The flowers are bright yellow, as pictured, in the morning, apricot in the middle of the day, and as they get ready to drop at nightfall are deep red.

### Coral Hibiscus

• Philippines: Araña, Spider Gumaméla
• Hawaii: Aloalo Ko 'ako 'a

*Hibiscus schizopetalus*

This dainty little coral colored hibiscus with a slender and graceful curving stem, adorns many topical gardens. The petals are frilled and lacy, and the center stamen droops like a pendulum on a clock.

## HOYA, Wax Vine, Wax Flower

• Fiji: Bitabita, Wa Tabua, Wabi, Draubibi • Hawaii: Pua-hoku-hihi • Malaysia: Akar setebal, Akar serapat • Samoa: Fue-sa

*Hoya carnosa*

This vine from China and Australia has thick shiny attractive leaves, and waxy, very fragrant star shaped white flowers with pink centers. It is a relative of the Stephanotis (see *Hargreaves* "Hawaii Blossom" Page 34). Hoya grows wild in Hong Kong, and was introduced into Fiji in 1880's. There are 24 native species in the Philippines. Cuttings of this vine make excellent house plants.

## IXORA, Jungle Flame, Flame of the Wood

- Ceylon: Ratmal, Rat-Ambala
- Fiji: Bulidavui, Suanibu, Sulu
- Malaysia: Pechah periok • Philippines: Santan, Santan-Pula (Tag.)
- Samoa: Suni, Filofiloa • Thailand: Khem

*Ixora macrothyrsa*

This member of the Coffee Family blooms most of the year. The big, bright round "snowball" heads of scarlet blossoms are very showy. The tiny small individual flowers that make up these heads have four petals.

The leaves, too, are glossy and handsome. The flowers and bark are used medicinally. There are many colors which are grown in hedges and shrubs throughout S. E. Asia and the South Pacific.

# JADE VINE

• Philippines: Bayou (Negritoes), Tayabac (Tagalog)

*Strongylodon macrobotrys*

The flowers on this vigorous Legume vine from the Philippines resemble jagged shark's teeth. They have been flourishing in the forests of Luzon and Mindanao since 1854. Now it is cultivated in most tropical areas because of the beautiful jade color of its 2″ to $3\frac{1}{2}$″ cornicopia-shaped blossoms that cascade downward for several feet. The huge green fruit houses 6 to 12 seeds that are shaped like a Brazil Nut. The flowers are made into exotic leis in Hawaii. Can be seen in the Royal Botanic Gardens in Ceylon, Bogor Gardens in Indonesia, and the Singapore Botanic Gardens.

## RED JADE VINE, Red New Guinea Creeper

*Mucuna bennettii*

This large climber from New Guinea has beautiful brilliant red flowers similar to the (green) Jade Vine.

The 3″ flowers of this vine, too, are like upturned beaks. They also have tri-foliate leaves.

Can be seen in a lovely big array in the Singapore Botanic Gardens, Ceylon Royal Botanic Gardens, New Caledonia, and The Philippines.

## JASMINE, Arabian Jasmin.

• Ceylon: Pitchia Mal • Guam: Sampagita, Sampaquita • Hawaii: Pikake • Malaysia: Melati, Melor • Philippines: Sampaguita, Campopot (Tag.) Sampag Kampopot • Samoa: Pua-solo-solo Society Islands: Tia Tia, Tafife • Thailand: Mali • Burma: sa-pai

*Jasminum sambac*

These are the National Flowers of the Philippines and Indonesia. They are tiny white, very sweet smelling, blossoms with 9 or 10 petals that look like a tiny rose. The leaves are small and shiny. The small bush belongs to the Olive Family from India. The flowers are picked when buds and strung into leis and garlands in Hawaii, Thailand and many other South Pacific Islands. Also used for commercial perfume.

## JATROPHA, Rose Flowered Jatropha

• Philippines: Chaya

*Jatropha hastata* syn.
*J. pandurifolia*

A small shrub that flowers continually has tiny bright red flowers, and is much used now for landscaping in gardens. It has been recently introduced into the Philippines, where it is called Miss Hong Kong. The leaves are sometimes ovate and sometimes fiddleshaped. Seen in most countries of S. E. Asia, Indonesia, Philippines, Hawaii.

## Gout Plant

*Jatropha podagrica*

Another Jatropha that has become popular for landscaping because of its gout-like stems that are swollen at their base, and coral-like orange red blossoms.

## LIPSTICK PLANT, Annatto, Arnotto

• Fiji: Kesa, Qisa, Ai Qisa • Malaysia: Kesumba • French: Roucou, Roucouyer, Achiot • Guam: Achiote • Hawaii: 'Alaea • Marquesas: Mahiha • Philippines: Achuete • Samoa: Loa

*Bixa orellana*

Ancient warriors used this plant (Annatto Dye) to paint their bodies the orange-red of the seeds. The people in Tonga and Samoa still use them to color their Tapa Cloth.

The dye has also been used to color cheese, margarine, candy, chocolate, cloth, lips, soap and paint. The whiskered pods make interesting dried flower arrangements.

The shrub has single pink 2″ flowers that resemble a wild rose. They grow in Micronesia, the Philippines, Malaysia.

## LOTUS, Indian Lotus, Sacred Lotus

• Ceylon: Nelun • China: Ho • Malaysia: Bunga Padam • Philippines: Baino (Tag.) • Thailand: Boa Luang, Dok Boa • Burma: Kyar-pan

*Nelumbium* (*Nelumbo*) *nucifera*

The Lotus is a beautiful plant of the Water Lily Family that was introduced to Egypt in 500 B.C. It is the National Emblem of Egypt, the Sacred Lotus of India, and introduced in prehistoric times in Asia.

It is the symbol of Buddhism. Buddha (and Buddhist saints) are pictured seated on the Lotus flower. The round wheel-like blossom is symbolic of perpetual cycles of existence. In China it is the symbol of purity and truth. (Also of summer). In India it is said that the Lotus was dyed with the blood of Siva, who was wounded by Kama, the Cupid of India. The Lotus is one of the few plants that has been cultivated since ancient times for its flower. It grows in ponds with 1′ to 3′ leaves like up-side-down umbrellas. The lovely 10″ rose colored fragrant flowers stay open for two days, but close at night. The fruit is a curious disk shaped flat receptacle with holes, each containing a seed—good for dried arrangements.

The roots, 2′ to 4′ tubers, grow in the mud at the bottom of the pond, and are used as a starch and vegetable. The sap is used for medicine, the fibers from the stems for lamp wicks, the seeds are nutritious and taste something like hominy.

**MELASTOMA** (The Melastoma Family is large)

*Melastoma sanguineum*

This is a small shrub with 3″ to 4″ pretty flowers of 5 to 8 purplish-pink petals, shiny leaves, deeply veined with silky hairs. Found in abundance at the sides of the road on the way up Peak Mt., Hong Kong.

## MELASTOMA (continued)

## PRINCESS FLOWER, Glory Bush, Lasiandra

*Tibouchina semidecandra*

A similar Melastoma grows wild in Hawaii. It is a Brazilian shrub whose 3″ flowers are deep purple or brilliant violet. The buds are rose red. It, too, has hairy stems and leaves. (Picture right is a similar Tibouchina from Australia.) In Fiji there is another Melastoma, *Medinilla waterhousei*, called Tagimaucia, Tekiteki Vuina, Moceawa by the Fijians. It is a beautiful plant with bright red bracts out of which peep the white flowers. It is one of Fiji's most distinctive flowers.

## STRAITS RHODODENDRON, Malabar Melastome, Singapore Rhododendron, Indian Rhododendron

- Fiji: Kaunisiga, Deriderikilotu, Dorodorokilotu, Sigadradrawa
- Malaysia: Sendudok

*M. malabathricum*

In Malaysia, these pretty purplish-pink 2″ to 4″ flowers grow all along the roadsides near Kuala Lumpur. The berry-like fruit contains seeds coated with a sweet red edible pulp which the birds are fond of and thus scatter the seeds. The sour young leaves are eaten in Java. The plant has astringent medicinal value, and the fruits, roots, and leaves contain a pink dye.

## MEXICAN CREEPER, Coral Vine

• Fiji: Corallita • Guam, Philippines: Cadena de Amor (Chain of Love)
• Thailand: Puang-chom-poo

*Antigonon leptopus*

Found along the banks, on fences, and clambering beside the road is this lacy, bright pink flower which suggests strings of small pink hearts. The flowers that have no petals are calyx with five petal like sepals. The leaves also are heart shaped with wavy margins. The vine belongs to the Buckwheat Family, and is beloved by the bees.

## MICKEY MOUSE PLANT

*Ochna kirkii*

This small shrub has pretty yellow flowers that cluster at the branch tips, but the fruits of this plant are the most interesting. Some think they look like a minature Mickey Mouse, others think of an elf or the "little people", because the red calyx has black seed cases that leave quizzical little faces when they fall off. Can be seen in Singapore Botanic Gardens and Ceylon Royal Botanic Gardens.

## MUSSAENDA, Ashanti Blood

*Mussaenda erythrophylla*

- Fiji: Bovu, Vobo, Vakacaredavu
- Samoa: Alo-alo-vao
- Philippines: Doña Trining
- Thailand: Don-ya

This bright African shrub is another member of the Coffee Family. Named after the late Pres. Roxas' wife in the Philippines, it is a striking sight with its small, star-like white flower. The hybrids below developed by Univ. of Philippines.

*M. sirikit* (after Queen of Thailand) *M. alicia*

## MUSSAENDA, Buddha's Lamp, Virgin Tree

• Malaysia: Balik adap • Philippines: Doña Aurora, Kahoy-Dlaga (Tag.) (Named after President Quezon's wife)

*Mussaenda philippica*
*var. Aurorae*

This small shrub is very popular in its native Philippines. It is also seen in Guam, Tonga, Fiji, Hong Kong, Singapore (Bot. Gardens), and Thailand. In Hong Kong it is called "Buddha's Lamp" because the tiny yellow flower looks like an oil lamp. This is the true flower, the white leaves are enlarged calyx lobes.

## BEACH NAUPAKA, Half Flower

• Ceylon: Takkada, Taccada Pith • Fiji: Dredre, Kativari, Kirakira, Vevedu • Guam: Nonasu • Hawaii: Naupaka-Kahakai

*Scaevola taccada* var sericea

Everywhere in the tropics this 3′ to 10′ spreading succulent shrub can be found growing wild on the beaches. There are many varieties. The small white flowers, streaked with purple, are odd because they look like just half a flower. According to Polynesian legend, lovers were separated leaving a half a flower of the youth blooming alone in the mountains, and the girl blossoming alone on the beach.

The berries turn white when ripe. Hawaiians sometimes call the plant "Huahekili," meaning hailstones because of these round white berries..called "Siale Tafa" in Tonga. The plant is used medicinally in Malaysia and Ceylon..

## MOUNTAIN NAUPAKA



• Hawaii: Naupaka-Kuahiwi

## NEEDLE FLOWER, Needle Flower Tree
## Tree Jasmine

*Posoqueria latifolia*

This small evergreen shrub has very fragrant 6″ white flowers which cluster at the branch tips. The edible fruit is yellow, and a little larger than a cherry. The 6″ long leathery leaves are deeply veined. Found in the Philippines, Royal Botanic Gardens in Ceylon, Bogar Gardens, Indonesia.

## OLEANDER

• Bali: Kenjari • French: Lauier • Hawaii: Oliwa, Oleana • Philippines: Adelfa • Thailand: Yee-toh • Tonga: Lokie

*Nerium oleander*

This colorful shrub is of the Periwinkle Family from Asia Minor. The leaves are slender, pointed, and a dull green. The branches are tipped with clusters of flowers of all colors, some single with five petals, others double. The colors range from white through cream, pink, rose, red, coral. The shrub is poisonous. Even insects do not bother it. Food cooked on the wood can even poison. Blooms continuously.

## ORCHIDS

*Books* have been written on orchids alone. In the Philippines, a similar vanda to the hybrid above, *Vanda sanderiana* is called "Waling Waling." It is brown and purple. The Malaysian word for orchid is "Anggerek"

*Spathoglottis plicata*

The Wesak Orchid *Denrobium macarthial* is endemic to Ceylon and is associated with the birth of Buddha. A festival celebrated when the moon is full is known as the Wesak Festival of Lanterns. Hong Kong, New Caledonia, Malaysia, The Philippines and Hawaii have many wild orchids that grow in the trees, the slopes and woods.

## GIANT ORCHID, Letter Plant

*Grammatophyllum speciosu*

The world's largest orchid plant is a native of Malaysia and the Philippines. The stems are 6′ to 12′, the flowers, that bloom on long stems 5′ to 7′ tall, are about 5″ across. They are an ochre-yellow with reddish-brown spots. They bloom on many stocks that rise from near the base of the huge plant. A fine plant can be seen in the Royal Botanic Gardens, Peradeniya in Ceylon. Seen also in Java, the Philippines, Fiji, and The Waterfall Gardens, Penang.

## ORCHIDS (continued)
## VANILLA

Tahiti: Vanille

*Vanilla fragrans* syn.
*V. planifolia*

Vanilla is the only orchid used commercially for food. It is a climbing vine with thick, oval, short-stemmed leaves, clinging aerial roots, and 2″ funnel-shaped, pale green flowers. These sometimes have to be hand pollinated to produce the 5″ to 6″ narrow pods. They are picked green, cured and left to ferment for 3 to 5 weeks then Vanillin (active part of Vanilla) crystallizes on the outside of the pod. About 100 lbs. of cured pods, called Vanilla Beans, are yielded to an acre.

The orchid blossoms 1½ yrs. after planting. It then bears thereafter for 30 to 40 yrs.

Grown commercially in Tahiti. The Vavau Islands of Tonga export Vanilla to Australia. It was introduced to Ceylon in 1847. Some grown in the Philippines and Hawaii.

## PAGODA FLOWER

• Hawaii: Lau ʻawa • Philippines: Biniuang (Tag.) • Thailand: Khem-shuti

*Clerodendron speciosissimum*
syn. *C. squamatum*

A rich blending of fuchsia-red tones distinguish this native of Java. They bloom in large, rather loose, upright heads. The individual flowers are five narrow lobes which turn back against the tube. The stamens and pistils curve beyond this flower in small red tufts. The star-shaped calyx may thicken and enclose the interesting, small, shiny, berry-like fruit (see above). The leaves are large, heart shaped, thick and velvety with deep veins. The stems are downy. There is a bright red variety (*C. rojo*) also.

## BANANA PASSION FLOWER, Granadilla, Giant Granadilla

• Ceylon: Rata Puhul, Seemaisora-kai • Hawaii: Lilikoi • Philippines: Kasaflora, Parola

*Passiflora quadrangularis*

This strong, quick growing climber is one of the 40 (out of 400 known species) of the Passiflora Family that is a native of Asia, South Pacific, and Madagascar. It has lovely purplish-pink 5″ flowers with greenish yellow 6″ to 10″ edible, three-grooved fruits. The flower, like all Passion Flowers, has the same center arrangement of stamens and styles that makes the family the symbol of Crucifixion, because early travelers thought it suggested the passion or suffering of Christ.

## PEPPER

• Ceylon: Gammiris, Molavu

*Piper nigrum*

The use of pepper as a spice dates back to *very* early times. The Pepper that we all know grows on a creeping perennial vine. Both black and white pepper comes from the same plant. The berries, or peppercorns, are picked when red (see picture), and spread on mats to dry in the sun. They then become black and shrivelled. When ground with this outer covering on, they are "black pepper". Without this covering they are "white pepper".

Pepper is an important crop in Sarawak, Malaysia. Also raised in the Philippines, Ceylon, India, Malaysia, Java. Black pepper is called "Lampong;" white "Muntok" in Ceylon.

## TEA

• Bali: Teh • Ceylon: Tá • Niue: Puka Puka *Thea sinensis*

The Tea Plant is an evergreen shrub that is a native of Assam, India. The short stemmed narrow oval and leathery 2″ by 5″ leaves contain oil glands. It has fragrant 1″ (usually 5 petaled) white or pinkish flowers. The fruit is a 3 parted, shiny, hard-shelled capsule with one seed. The vigorous new shoots that develop after plant is pruned are the source of the commercial product. Green tea comes from heating and rolling leaves. Black tea comes from heating, rolling, and fermenting the leaves. Oil and alkaloid like caffeine gives tea its flavor. Long brewing extracts bitter tannin. It was a drink used medicinally originally, but since the 5th century, it has been the chief beverage in China, and in Europe since the 17th century. Teas are scented by adding Jasmine, or other flowers, and orange to drying leaves.

## PETREA, Sand Paper Vine, Purple Wreath

*Petrea volubilis*

This vine bears profuse sprays of lovely bluish-violet flowers which cascade from its branches. The sepals are lighter in color than the corolla and outlast them on the vine leaving it a dull grey color. These sepals are the wings for the seeds. The leaves have a harsh rough surface giving the vine the name "Sand Paper". There is also a white variety called Bridal Wreath (*P. kohautiana*)—see picture at right. Can be seen in the Penang Waterfall Gardens in Malaysia, Royal Botanic Gardens in Ceylon, Bogor Gardens in Indonesia.

## PHILIPPINE VIOLET

Philippines: Violetas

*Barleria cristata*

A small 2′ to 4′ shrub, native of India, often grown for hedges and borders. The 2″ flowers look like a violet and are a deep violet color varying sometimes from a lighter color to white and pale pink. They are stemless with a four-parted calyx and funnel shaped five-lobed, 2″ corolla. Seen in the Philippines, Guam, Malaysia, Hawaii.

## DWARF POINCIANA, Peacock Flower, Pride of Barbados.

• Hawaii: Ohai Alii • Philippines, Guam: Caballero • Malaysia: Sepang • Marquesas: Kohai • Rarotonga: Tartara Moa • Niue: Fiti Hetau • Samoa: Lau-pa

*Poinciana pulcherrima* syn. *Caesalpinia pulcherrima*

Bright clusters of fiery red (also yellow variety) flowers grow on the tips of this small graceful shrub or small tree. This is not a true Poinciana, but a close relative. It has five petals which are margined by yellow. There are long stamens and a pistil projecting from the center of each butterfly-like flower. It has lacy foliage with prickly branches, and is a member of the Legume Family. It is the sacred flower of Siva in India.

## WILD POINSETTIA

• Ceylon: Chaconia

**W***arscewiczia coccinea*

This interesting tropical America shrub produces large sprays of vermilion inflorescence. Much like its namesake, the Poinsettia, the showy parts of these plants are not the flowers, but the enlarged red sepals. Introduced into Ceylon in 1923.

## POWDERPUFF LILY, Blood Lily, Sea Egg

*Haemanthus multiflorus*

These beautiful 6″ to 8″ red powderpuff balls grow from a large bulb. The flower head consists of 50 to 100 tiny red flowers with red stamens which give it a lacy appearance. They continue blooming a week to 10 days. The plant originated in Africa, but thrives in the Philippines, Hawaii and other Pacific areas. Also seen in Guam and Ceylon.

# RANGOON CREEPER

*Quisqualis indica*

This large woody vine from S. E. Asia, Malaysia, and the Philippines has attractive, fragrant clusters of light pink to deep red 3″ narrow tube flowers. The fruit is 1″ ovoid and sometimes used medicinally in Malaysia along with the roots and leaves. Found also on Guam, Ceylon and Fiji.

## RICE

- Bali: Bras (white rice), Gaga (red rice) Indjin (black rice) • Fiji: Dahn • Hawaii: Laiki • Malaysia: Padi, Paddy

*Oryza sativa*

Rice is the main food of more than half of the people in the world. It is a 3′ swamp grass with flat 10″ leaves and flowers 6″ to 12″ that bloom at the ends of the stalks. When the grain begins to ripen, and these panicles droop with their weight, the water (see insert above) is drained from the "Paddy" to hasten the harvest. Balinese, who have two crops a year, are the best rice growers in the Indonesian Archipelago. As above, their fields are beautiful to see. The Philippines have three crops a year, and are continually experimenting at the University of the Philippines, as is the Department of Agriculture in Fiji conducting extensive trials. Rice was introduced to Fiji in about 1902. Hong Kong grows rice for its own consumption. Some rice is grown in Ceylon, Malaysia, and Thailand. Other than food, rice is used as a laundry starch, fermented into liquor; Sake in Japan, Mekong in Thailand, Arak, a distilled brandy, and Brom, a sherry made from black rice. The straw of the rice plant is used for clothing and packing.

## SHRIMP PLANT

- Fiji: Honolulu Salvia
- New Zealand: Lobster Plant

*Beloperone guttata*

The heart shaped bracts on these Acanthus Family leggy plants suggest a scale that looks like the curved tail of the shrimp. The true flowers are tiny white and tubular with purplish veins that peek out from the lower lip of two lobes. The yellow variety at right was found in a small botanic garden in Tonga.

# SILVERSWORD

• Hawaii: Hinahina, 'Ahinahina

*Argyroxiphium sandwicense*

This startling Hawaiian plant is endemic to the high mountains of Hawaii. It is found only on about the 10,000 foot level of Haleakala, East Maui, and the 6,000′ to 12,000′ mountains of Hawaii, the big island. It grows among the volcanic rocks. First forming is the lovely round silver 2′ rosette (above), then 9 to 14 years later, when the plant has reached maturity, a single 6′ specacular swordlike bloom with many 2″ flower heads blooming all along the stalk, thrusts out from its center. Black fruits are then formed and the plant dies.

# TAPIOCA, Cassava

• Bali: Ketela • Ceylon: Manjoca • Fiji: Cassawa, Kasava, Yabia ni Vavalagi, Kasera • Java, Malaysia: Ubi kayki, Mandiocca • New Caledonia: Manioc • Philippines: Kamoting Kahoi • Samoa, Tahiti, Tonga: Manioka, Maniota, Lapioka, Ufi-la 'au, Ufi-maniota

*Manihot esculenta* syn.
*Manihot utilissima*

Tapioca has been a source of food since early times. This 3′ to 9′ bush is grown for its long, edible, tuberous roots that resemble sweet potatoes. They are high in starch. The sap is poisonous, but this disappears on cooking and washing. After boiling these roots, the starch Tapioca can be extracted by grating and passing through fine mesh. It is then heated to form the granulated Tapioca of commerce. The plant is cultivated widely in Bali, Fiji, New Caledonia, Malaysia, Tahiti, Ceylon and the Philippines. There is also an ornamental Cassava with green and white leaves. (right).

## THUNBERGIA, Sky Flower, Bengal Clockvine, Bengal Trumpet

*Thunbergia grandiflora*

One of the loveliest lavender-blues is this vine that grows in so many tropical gardens. It is often combined with the white Thunbergia or the yellow Allamanda (see page 3). It thrives practically anywhere, and blooms almost continually. The funnel shaped flowers of five lobes have a pale yellow throat. It is a native of India of the Acanthus Family.

## XANTHOSTEMON

*Xanthostemon rubrum*

This is one of the endemic plants of New Caledonia that grows on the ore (Terre rouge) grounds that cover the island's mountains. The lovely red flowers on this bush are surrounded by dark green leathery leaves. There are other species in New Guinea and Australia.

## Publications by the same authors:

All books in this family are 64 pages each, all have over 100 full color pictures (some as many as 130) and all are the same size and format. Local names in local languages and text are different to reflect the countries and geographical areas they cover. Botanical names are included.

- "TROPICAL BLOSSOMS of the CARIBBEAN"
- "TROPICAL TREES found in the CARIBBEAN, South America, Central America, Mexico"
- "AFRICAN BLOSSOMS" (covers Tropical Africa, South Africa, Madagascar, Mauritius)
- "AFRICAN TREES" (covers same areas as "African Blossoms")
- "HAWAII BLOSSOMS"
- "TROPICAL TREES of HAWAII"
- "TROPICAL BLOSSOMS of the PACIFIC" (covers S.E. Asia, Malaysia, Ceylon and Pacific Ocean countries)
- "TROPICAL TREES of the PACIFIC" (covers same areas as "Tropical Blossoms of the Pacific")

All books can generally be found in book stores and tourist shops in the particular countries they cover. Or, books will be mailed postage paid via surface mail anywhere in the world for $2.50 each in U.S. funds from the publisher:

HARGREAVES COMPANY, INC.
Box 895, Kailua, Hawaii, 96734, U.S.A.

IMPORTANT: If AIR MAIL delivery *outside* the U.S.A. or territories is desired, add $1.80 U.S. for one copy, plus 60¢ for each additional copy to cover extra airmail postage. If AIR MAIL delivery *inside* the U.S. or territories is desired, add $1.50 for up to two copies, plus 50¢ for each additional two copies.
When ordering from outside the U.S.A. please send payment with order in U.S. currency, International Postal Money Order, Bank Draft, or check on any U.S. bank.

Lithographed in Japan